Six Sigma Series

DFSS – Design for Six Sigma

Klaus Hogreve

DEDICATION

This print version on my book is dedicated to the few old fashioned people who still prefer a print version over a digital download. Enjoy!

CONTENTS

Acknowledgments   i

1   Preface   1

2   Design for Six Sigma (DFSS)   3

3   QFD (Quality Function Development)   11

4   FMEA (Failure Modes and Effects Analysis)   18

5   DOE (Design of Experience)   27

6   RDO (Robust Design Optimization)   31

Appendix A   Simple Lean Design tools   34

Appendix B   Innovation   37

Appendix C   Severity Rating Scale   39

Appendix D   Occurrence Rating Scale   40

Appendix E   Detection Rating Scale   42

Appendix F   Sample FMEA Worksheet   43

ACKNOWLEDGMENTS

Special thanks to Jürgen Blankenburg for providing me with a vast amount of Six Sigma project information and to Kamran Avary, PhD, who inspired me to write the Six Sigma Series.

Special thanks to Yasmin Hogreve for the cover design.

1. PERFACE

In the current market place there is a high demand of new products since the product life cycles are getting shorter and shorter. The organizations success depends on its ability to introduce new and successful products on a fast rate. However, new product failure rates are still between 40% and 90%, depending on the industry and the effort the company puts into marketed research.

On top of this, about 80% of all quality problems are design related. And I am not talking about the intentional failures integrated into a product to shorten its life span. Therefore a well thought thru and tested design will avoid quality problems down the road.

Design for Six Sigma (DFSS) is a response to the demand of faster design of new products while radically improving the success rate and quality of these products.

DFSS is a product design process in which the focus is on the "voice of the customer". The shortcoming of Six Sigma is that even if you have a "zero defects" process, what is it worth when the result of all the effort is useless for the customer? DFSS is trying to design products and services which the customer really wants. "If you build it they will come" doesn't work anymore. At least not in the long range. DFSS can be seen as a tool of Strategic Marketing.

The other situation where to use DFSS is when Six Sigma projects did not fix the problem in an already established process and did not result in the expected improvement. This means that there is a flaw in the process which asks for a complete redesign of the process.

DFSS seems to me to be the step child of the process improvement movement. Also some call it the future if Six Sigma. It is a fascinating tool, but it does not have the place in the literature it deserves. Usually it gets about one chapter at the end of most Six Sigma books. I am here to change this and devote a whole book to DFSS.

Prominent DFSS tools are QFD (Quality Function Deployment) and FMEA (Failure Modes and Effects Analysis).

2. DESIGN FOR SIX SIGMA (DFSS)

In the current business climate if is essential for an organization in order to be successful to come up with new products on an ever escalating rate. For example: Up until the early 80s Mercedes Benz run a very successful car business on only 3 different platforms and very few variations: There was the mid size sedan which became the E class, there was the luxury sedan which became the S class, and there was the roaster (SL/SLC). The mid size and luxury sedan lineup got a quantum leap do-over about every 8 years; the roasters were even on a longer cycle (the R107 run for almost 20 years). BMW was in a similar situation with its 02 series and the 2500 / 3.0. If we look at these companies now, this has changed dramatically. New products are now coming on a way faster scale and the customers now request this.

For several years the executive management of almost every company has set goals like x% of revenue to come from new products, leaving the burden of innovation to the middle management.

Given a product failure rate of 40% to 90% depending on industry and effort of customer research data been used the product development, the promise of reducing product failure rate and shorten product development times of DFSS seems to be perfect solution. On top of this, about 80% of the quality problems are design related. That gives a thought thru and tested product and service design the benefit of reduced cost down the road.

The objective of DFSS is to create products and processes that are immune to problems, delight customers, and inspire confidence in a company's ability to deliver value and excellence. But there is also a downside: Implementing DFSS requires a significant layout of time and money. There is no quick ROI (Return On Investment); making DFSS a strategic tool with long term benefits. But those long term benefits will put the organization in the position of strength in the long run.

The benefits of DFSS are:
- Ensures that the VOC (Voice Of the Customer) is included in each stage of the product design, reducing the product "flop" rate.
- Integration of various functions in the design process, reducing miscommunication that leads to rework and other waist.
- It helps in creating lower defect rates when process improvements (Six Sigma) failed to reach desired levels (redesign of processes).

2.1. SIX SIGMA VERSUS DFSS

Both, Six Sigma and DFSS have some elements in common:

Strategy	Both improve product and service quality.
Defects	Both reduce defects.
Effects on costs and cycle time	Both lower costs and reduce cycle time.
Process	Both employ a systematic, toll-gate process.
Tools	Both use similar statistical and quality tools.
Approach	Both carry out data-driven and fact-driven analyses.
Teams	Both work through teams formed of Master Black Belts, Black Belts, Green Belts, a process owner, and need a Champion.

The differences between both are that:

- Six Sigma is **reactive** to problems, removing defects and DFSS is **proactive**, avoiding the problems in the first place.
- Six Sigma aims for fast returns (immediate results) and DFSS is a long term strategy with payout way down the road.
- Six Sigma focuses on improvement projects and DFSS focuses on products and processes.
- Six Sigma improves the current process and DFSS optimizes new designs and processes.

The Six Sigma uses the DMAIC methodology while DFSS uses the DMAVD or the ODOV metrology.

2.2. DMAIV (DEFINE, MEASURE, ANALYZE, IMPROVE, VERIFY)

The DMAIC project methodology has five phases:

- *Define* the system, the voice of the customer and their requirements, and the project goals, specifically.
- *Measure* key aspects of the current process and collect relevant data.
- *Analyze* the data to investigate and verify cause-and-effect relationships. Determine what the relationships are, and attempt to ensure that all factors have been considered. Seek out the root cause of the defect under investigation.
- *Improve* or optimize the current process based upon data analysis using techniques such as design of experiments, poka yoke (mistake proofing), and standard work to create a new, future state process. Set up pilot runs to establish process capability.
- *Control* the future state process to ensure that any deviations from target are corrected before they result in defects. Implement control systems such as statistical process control, production boards, visual workplaces, and continuously monitor the process.

2.3. DMADV (DEFINE, MEASURE, ANALYZE, DESIGN, VERIFY)

DMADV

The DMADV project methodology, known as DFSS ("**D**esign **F**or **S**ix **S**igma"), features also five phases:
- *Define* design goals that are consistent with customer demands and the enterprise strategy.
- *Measure* and identify CTQs (characteristics that are **C**ritical **T**o **Q**uality), product capabilities, production process capability, and risks. = Voice of the Customer
- *Analyze* to develop and design alternatives.
- *Design* an improved alternative, best suited solution per analysis in the previous step
- *Verify* the design, set up pilot runs, implement the production process and hand it over to the process owner(s).

2.4. IDOV (IDENTIFY, DESIGN, OPTIMIZE, VERITY)

The IDOV project metrology, the alternative DFSS method, contains of these four phases:
- *Identify* the design goals by incorporating the customer requirements into the formal product design. Cross-functional teams assign responsibilities and define the customer requirements or voice of the customer (VOC) by conducting a competitive analysis and defining the critical-to-quality (CTQ) parameters. Common tools in this phase are QFD, FMEA, and benchmarking.
- *Design* the product, process, or service by developing various concept designs. These design alternatives are evaluated, and the best-fit concept is selected from them. Then, based on this concept design, the raw materials, process or service scope, procurement, and development plans are developed.
- *Optimize* the product, process, or service by establishing product specifications, service parameters, and process settings. The process capability information is used to predict performance and optimize the designs where possible. This information is also used to build error proofing into the manufacturing or service process in order to reduce variation, costs, and defects.
- *Verify* the results by doing test runs and evaluate prototype designs in order to test and validate the design and quality control systems.

The DMADV and IDOV methodologies ensure that design development is done by cross-functional teams. These methodologies identify customer needs and conformance to specifications before starting a project. They explore customer quality requirements as well as product or delivery quality requirements. They also focus on early detection and rectification of errors, therefore ruling out much of the rework associated with traditional measures.

2.5. IS DFSS ALWAYS ADVISED?

DFSS is not a cure for all problems. And since it requires a substantial upfront investment, it is strongly advised to verify that DFSS will the best toolset available and the projected results will justify the investment. As I like to say: A hammer is a great tool, but sometimes a wrench or a screwdriver will do a better job.

In order to decide if to deploy DFSS, these four things should be considered:
- The Six Sigma level of the organization's development: If the level is rising steadily, it is unlikely that DFSS will be beneficial, if the rise of the level is slowing significantly or is still low, DFSS is likely to be beneficial.
- The organizations business environment: If changes in market demand, customer requirements, technology, or legal restrictions threaten to make the products or services obsolete, DFSS can help the organization to adapt.
- The organization's prioritized project schedule: Simple projects be handled first, and the organization should be prepared to roll out DFSS for the more complex projects.
- The organization's capacity to roll out the project: Does the organization have the resources needed to roll out a DFSS project? Does it have the financial capability? And even more important: Is the team and management available to be involved?

There are three main reasons to deploy DFSS:
- A new product, service, or business is currently not in place, and its design or processes need to be designed from scratch.
- All possible efforts have been made to meet customer expectations, but customer satisfaction is still too low.
- After many Six Sigma projects, the rate of improvement has slowed or stopped and the organization is still not close to Six Sigma level.

As shown above, DFSS is not always the best solution. In order to avoid an investment into the wrong tools, it is important to verify its feasibility and benefits first.

If a process is not up to the standard, a Six Sigma Project or just the application of some Lean tools may fix the problem faster and more cost effective, but if these tools have failed and the organization is ready to scrap the existing process, DFSS may be the answer.

3. QFD (QUALITY FUNCTION DEPLOYMENT)

In order to be able to compete in a global market with increasingly sophisticated customers, successful businesses have shifted their attention from the product to the customer, enabling them to respond faster and more effective to changes in the marketplace. However, to enable fast and effective response to changes in the market, the businesses must have a crystal clear understanding of the customer's wants, needs, as well as their perception of the products and services and their features. The process to archive this is called Quality Function Deployment (QFD) and starts with including the Voice of the Customer (VOC) in every stage of the business process from concept to delivery. To archive this, the VOC is transferred into the House of Quality (HOQ) Matrix, enabling the design team to create the right products and services.

Quality, in this context, is not only meeting the promises made to the customer and delivering what the customer expects; it aims to identify the customer requirements that have a positive effect of perceived quality and ensure a high rate of customer satisfaction.

3.1. VOC (VOICE OF THE CUSTOMER)

In order to get a crystal clear understanding of the customer's wants, needs, as well as their perception of the products and services and their features, a business must go beyond simply asking the customer what is expected. To ensure the success of customer-focused strategies, the customer requirements are often categorized using a Kano analysis. These three Kano categories are:

- **Dissatisfiers:** Dissatisfiers are basic requirements in a product or service that cause your customer not to like the product or service if they are missing.
- **Satisfiers:** Satisfiers are additional features whose presence in the product or service satisfies the customer. These features increase levels of customer satisfaction, resulting in a product or service that meets customer requirements. Satisfiers aren't necessarily expected as a regular part of the package. But the more satisfiers there are, the more pleased the customer will be with the product or service.
- **Delighters:** Delighters are features that go beyond the customer's expectations – they are unexpected features that impress customers. The absence of delighters does not dissatisfy customers, but their presence serves to delight customers. Discovering features that cause customer delight without negatively affecting profitability is the key to gaining a competitive edge.

To use the example of a retirement home, the provision of 24 hour nursing staff, appropriate food, and rooms are dissatisfiers; their absents would create absolute dissatisfaction. Alternative therapies, attractive public areas, organized events, and a bus service to the mall or golf course will increase customer satisfaction and will qualify as satisfier. The ability to order meals from a menu and concierge service will be delighters.

The problem is that once a feature is classified in one category, it is not certain that it will remain there. Just as an example airline seats completely reclining into a flat bed - this feature was a delighter just a few years ago and has now become a dissatisfier. With this in mind, the future successful business will not rest once it has the "perfect" product or service; it will continuously verify with the customer that its product or service is still cutting edge. Almost as bad as not having the requested features is spending resources on features the customer does not care about.

3.2. HOQ (HOUSE OF QUALITY) MATRIX

The House of Quality (HOQ) Matrix transfers the customer requirements ("What's") into technical requirements ("How's") by transforming a list of prioritized customer requirements into a list of engineering targets and their target values that need to be met by the new product, service, or process.

The HOQ Matrix contains of six components:

- The "What's", the **Customer Requirements**, list the customer needs and wants (VOC) for a product, process, or service gathered from the customer surveys, interviews, e.g.
- The Prioritization of the "What's", the **Customer Importance**, states the customers' perception and prioritization of all of the listed Customer requirements as gathered by the market survey. The importance is listed on a linear scale, mostly between 1 and 5 or 1 and 10.
- The "How's", the **Technical Requirements**, lists the specifications of the product, process, or service as defined by the design team. They define how the customer requirements will be met and must be measurable and given target values.
- **The Interrelationship Matrix** component shows the design team's perception of the interrelationship between the customer requirements and the technical requirements. This relationship between the "What's" and 'How's" is usually classified as either strong, medium, or weak.
- The **Technical Correlation Matrix** identifies if the technical requirements (How's) support or impede each other in the design. This helps the designer to identify trade-offs needed to be made between the technical requirement elements. These technical correlation are usually classified as positive or strong positive if they support each other, or negative or strong negative if they impede or hinder each other.

- The **Technical Priorities** or priorities of the "How's" shows the priorities, measure of technical performance archived by competitive products, and the degree of difficulty involved in developing each technical requirement. The product design is evaluated in respect to these priorities during the design process. These technical priorities are usually rated on a scale from 1 to 5 or 1 to 10.

The HOQ matrix is used to document the design team's perception of the relationships between the different customer and technical requirements so that they can be used in the downstream processes. Typically this matrix is created by a cross-functional team able to provide technical and quality (as defined above) expertise.

Generally, the HOQ Matrix development begins with listing the Customer Requirement as provided by the VOC. In the next step the Technical Requirements will be developed and added to the matrix. Then the priorities for customer requirements will be set based on the VOC information. Now the relationship weighting of the customer requirements and the technical requirements will be rated in the Interrelationship Matrix, using a sale from 1 to 10. After the technical correlation matrix has been completed, the technical priorities are calculated by adding the relationship weight of each customer requirement for each technical requirement.

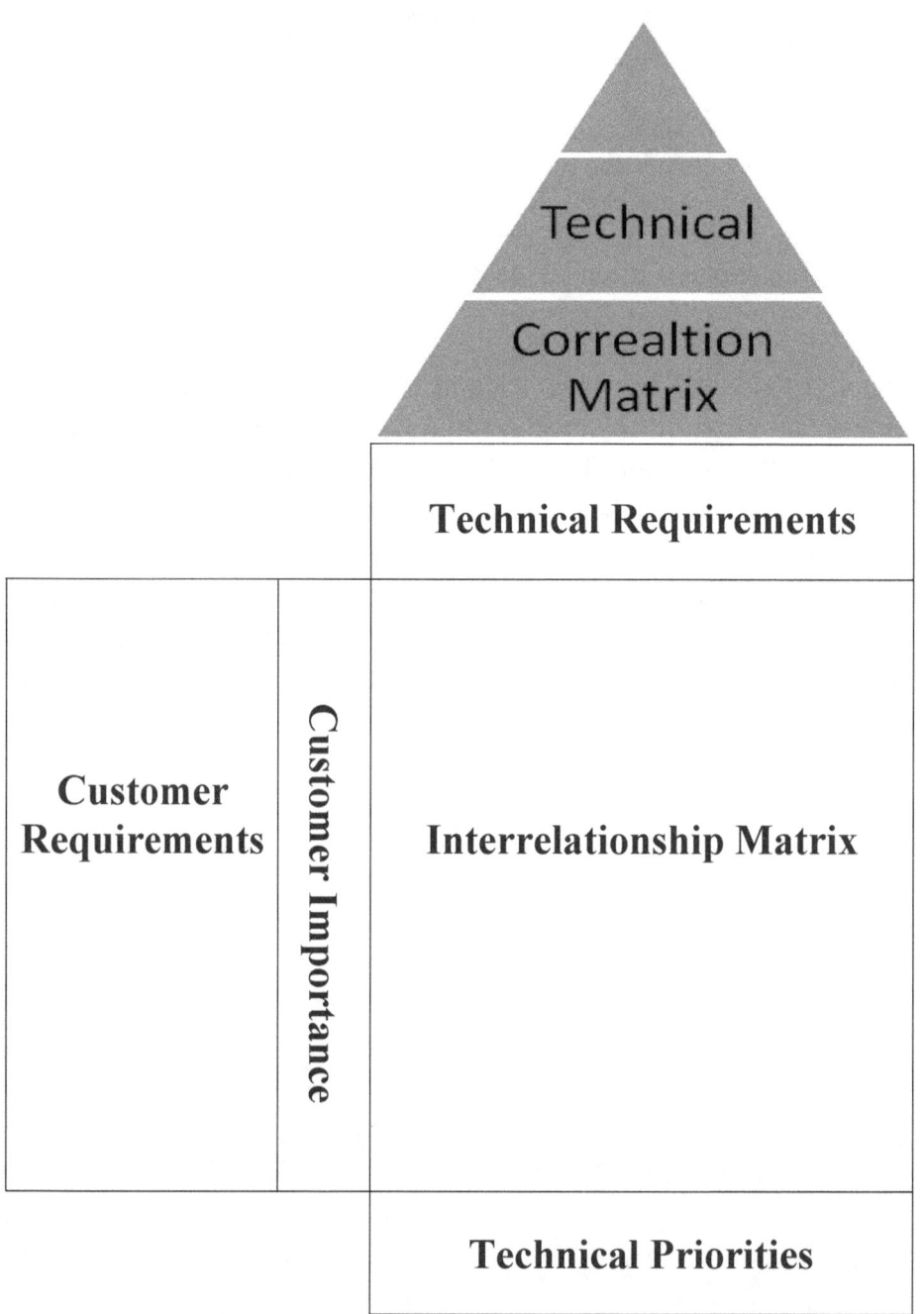

Phase I - Product Planning for a Cheese Burger		Importance to the Customer	Technical Requirements									Customer Satisfaction Index	Competitive Assessment	Sales Point (1.0, 1.2.., 1.5)	Needs Weight	Customer Assessment				
			Fresh			Tastes Good		Healthy												
			Bacteria count	Thawing time	Bun shelf life	# of spices	# of returns	Cheese slice diameter	% fat	# of organic suppliers	# of additives to meat					1	2	3	4	5
Customer Needs	Fresh	No bugs	5	◉	◉	◉		○		△		◉	200	1.0	1.0	5.0	⊖ ⊗ $ +
		Not frozen	2	○	◉	◉	△				◉		62.	1.0	1.2	2.4	⊗$⊖ +
	Tastes Good	Soft bun	3			◉		△		◉		◉	84.	4.0	1.2	14.	+ ⊖ $ ⊗
		Spicy meat	1				◉					◉	18.	3.0	1.0	3.0	⊗+$ ⊖
		Cooked my way	3		△			◉	○	△			42.	3.0	1.2	10.	$+⊖⊗
	Healthy	Lots of cheese	4						◉	◉		△	76.	4.0	1.5	24.	+ $⊖ ⊗
		Low fat	5		○		△		◉	◉			110	5.0	1.2	30.	+$⊗ ⊖
		Organic	2	○	○	○	△		△	○	◉	◉	64.	1.0	1.1	2.2	⊗$⊖ +
Target Values				.1	4 hrs.	3 days	4	3%	4 inches	98%	3	3								

Technical Assessment	Bacteria count	Thawing time	Bun shelf life	# of spices	# of returns	Cheese slice diameter	% fat	# of organic suppliers	# of additives to meat
+ Our Burgers 6 ⊗ Betty's Burgers 4 $ McBecker's 3 ⊖ Cow in a Bag 2 1	+ ⊖ ⊗ $	+ $ ⊖ ⊗	⊖ ⊗ ⊖ +	⊖ $ + ⊗	+ ⊖ ⊗ $	⊗ ⊖ $ +	⊖ ⊗ $ +	⊗ $⊖ +	+ ⊖ $⊗
Importance Rating	57	87	96	16	47	92	122	81	58

Above is an example for a HOQ Matrix for a Cheese Burger. This example also includes the analysis of the competition products, which helps to differentiate the new product from the competition.

4. FEMA (FAILURE MODES AND EFFECTS ANALYSIS)

Failure Modes and Effects Analysis (FMEA) is a proactive quality tool that helps to identify and prevent potential errors and to develop and implement corrective actions to address failures before they reach the customer. It's de facto a tool of risk management deployed in the early phases of design or process development. Therefore, FMEA enables the team to identify, define, prioritize, and eliminate known or potential failures of the design, system, or manufacturing process. It is an effective tool to minimize the risks associated with these failures or potential failures before they reach the customer, hopefully even before they design leave the development phase. It encourages critique of design and promotes dialog between departments to ensure all aspects are reviewed before the product or process reaches the development phase.

FMEA forces the team to quantify and prioritize risks associated with failure modes, and to calculate the probability, effect, and severity of potential failures. This tool leads to the prioritization of corrective actions and solutions to eliminate or to manage the failures and their effects. Just to set things straight; there is no perfect and foolproof product, service, or process; at least not within a range of economically justifiable solutions, but the next best thing to do is to manage the risks resulting from the shortcomings, resulting in a design what is still economically feasible but minimizes the weighted potential payout of the potential failures. Payout in this case means not only monetary payout due to lawsuits and settlements; it also includes the cost of loss of reputation and the cost of redesign, rework, recalls, e.g..

The high-level steps of implementing or using the FMEA tool are:

1.   **Select the area of focus** - When selecting the area of focus, team members decide on a single process, product, or service that they will analyze using the FMEA tools. Once one area is dealt with by step 2 – 7, the team moves on to the next area.

2.   **Study the focus area** - Team members from different departments such as design, manufacturing, marketing, and quality study the focus area – the identified process, product, or service – to gain a deep understanding of it and its context.

3.   **Identify potential failure modes** - Once the team members have studied the design, they identify potential failure modes. These may be caused by issues such as components, processes, materials, and costs.

4.   **Rate the levels** - Once the failure modes have been identified, the team members define and rate the levels according to their perceived levels of severity, occurrence, and detection.

5.   **Prioritize** - Once the severity, occurrence, and detection levels of potential failure modes have been assigned, a risk priority number (RPN) is assigned to each mode, thereby prioritizing failures.

6.   **Implement and evaluate** - Corrective actions addressing high priority failure modes are implemented and evaluated. The risk priority number is then recalculated for each mode.

7.   **Update the table** - Once corrective actions have been implemented and priorities have been reassigned, the person responsible for the project should then update the FMEA table as well as the design documents to ensure that the information is current.

Using the FMEA tool generates numerous benefits for a company:

•   improves the quality and reliability of designs, products, services, and processes

•   helps to identify critical-to-quality (CTQ) characteristics

•   reduces development time, cost, and problems by fixing the problems in early stages of the development process

•   documents and tracks risk reduction activities, which improves the corporate risk management

•   captures the collective knowledge of a team

The mainstream literature shows at least for FMEA types:

- **System FMEA** - A System FMEA is a sub-category of Design FMEAs. It is used to analyze system functionality in the first stages of design – for example, before specific hardware has been determined. This tool helps to identify potential failure modes associated with the functionality of the system that may be caused by the system's design.

- **Design FMEA** - A Design FMEA is used to analyze component designs during the development stages of design – for example, after specific hardware has been determined but before design drawings have been released. It focuses on potential component failures caused by design errors.

- **Service FMEA** - A Service FMEA is used to analyze services for potential failure modes caused by system or process problems. This is conducted before the first service reaches the customer. It includes non-manufacturing aspects such as financial and legal services, education, health care, and hospitality services.

- **Process FMEA** - A Process FMEA is used to analyze transactional processes, such as those in manufacturing and assembly, to identify failures caused by these processes. This type of FMEA typically occurs after a Design FMEA has been conducted.

As much of FMEA types one comes up with, they are all sub-categories of the two main FMEA types, which are **Design FMEA** (DFMEA) and **Process FMEA** (PFMEA).

PFMEAs differ from DFMEAs in terms of their focus and purpose but their individual steps are very similar. PFMEA requires a team to review design characteristics relative to the manufacturing process. Failures and corrective actions are focused on processes rather than design. DFMEA relies on product design changes to overcome process weakness.

4.1. DFMEA (DESIGN FMEA)

DFMEA is developed by a cross-functional team, generally comprising representatives from a variety of engineering functions including quality, product design, manufacturing, logistics, and testing. The team is ideally made up of five to ten members, who each analyze the product design. All of the team members then recommend design changes and follow through on recommended actions, before reassessing the design.

When analyzing a product design, the teams will focus first on prevention, then on early detection options, and finally on corrective and mitigation actions. DFMEA documents the weaknesses in the product design that may cause system failures, thereby helping to eliminate or reduce unsafe conditions that could result from these failures. The team is then able to modify the design as necessary and assess the revised failure modes to ensure a safe and reliable product.

DFMEA helps to improve quality and reduce waste, encouraging the following benefits for organizations:

- it identifies potential product failure modes early in the process
- it identifies characteristics that require special controls and highlights areas of improvement
- it ensures that potential failure modes and their effects are considered
- it evaluates product design requirements and testing methods
- it prioritizes design improvements
- it documents the rationale behind the design
- it improves the system's safety
- it reduces failure rates during the product life cycle and increases the duration of useful life cycles in products

The steps to conduct a DFMEA are:

1. Review the design – use the design documents like blueprints, models, or schematics of the design / product to identify each component and interface.
2. Brainstorm potential failure modes – review existing documentation and data for clues. If the design is in a computer design, simulate this use, see were the stress points are, and identify potential problems. List them on the FEMA worksheet.
3. List potential effects of failure – there may be multiple effects for each failure mode.
4. Assign severity rankings – based on the severity of the consequences of failure. See the provided ranking guide in Appendix C as a suggestion.
5. Assign occurrence ranking – based on the expected frequency this failure may occur. See Appendix D for suggestion of the rating.
6. Assign detection rankings – based on chances that the failure will be detected before the product reaches the customer. This is a ranking of the effectiveness of the detection control. See Appendix E for suggestions.
7. Calculate the RPN – severity * occurrence * detection
8. Develop the action plan – define a plan to fix the potential design failure. Define the action to be taken, by whom, and when.
9. Take action – implement the improvement outlined in step 8.
10. Reassess the severity, occurrence, and detection ranking based on the improvements and recalculate the RPN.
11. Re-evaluate if the RPN is acceptable or if additional improvements are needed.

4.2. PFMEA (PROCESS FMEA)

When an understanding of the processes needed for a design to become a product or service is reached, the team can begin a Process FMEA. PFMEA is an analytical technique that identifies potential product-related process failure modes and manufacturing or assembly process causes of failure, assesses the potential customer effects of the failures, and identifies variables to focus controls to prevent and detect failures. PFMEA can also assist in developing new machine or equipment processes.

The benefits of conducting a PFMEA include

- improvements in the detection and elimination of defects in products and processes
- assistance in the development of process control plans
- prioritization of improvement activities
- guidance in creating future improvement activity plans
- identification of Six Sigma projects
- documentation of the rationale behind changes to processes

As with DFMEA, a cross-functional team develops the PFMEA. The team comprises five to ten members – typically, representatives from design, manufacturing, and quality functions, who each analyze the process design. The team members then recommend changes to the processes and follow through on recommended actions before reassessing the process.

The steps to conduct a PFMEA are:
    1. Review the process – use the process flow chart or process map to identify each process component.

2. Brainstorm potential failure modes – review existing documentation and data for clues. List them on the FEMA worksheet.
3. List potential effects of failure – there may be multiple effects for each failure mode.
4. Assign severity rankings – based on the severity of the consequences of failure. The ranking guide in Appendix C is a suggestion / guideline.
5. Assign occurrence ranking – based on the expected frequency this failure may occur. See Appendix D for suggestion of the rating.
6. Assign detection rankings – based on chances that the failure will be detected before the product reaches the customer. This is a ranking of the effectiveness of the detection control. See Appendix E for suggestions.
7. Calculate the RPN – severity * occurrence * detection
8. Develop the action plan – define a plan to fix the potential process failure. Define the action to be taken, by whom, and when.
9. Take action – implement the improvement outlined in step 8.
10. Reassess the severity, occurrence, and detection ranking based on the improvements and recalculate the RPN.
11. Re-evaluate if the RPN is acceptable or if additional improvements for this specific area are needed.

4.3. FMEA WORKSHEET

The FMEA worksheet is basically the same form for DFMEA and PFMEA. It is a worksheet to document potential problems and corrective actions. The only difference is the type of project information that the worksheets contain.

It is up to the creativity of the team on how it will be laid out in specific, but they all should have two main sections, the Background or Problem section made up of the Description, Failure Mode, Effect and Cause (Root Cause) sub-sections with its ratings, and the Countermeasure section showing the action to be taken, the who and the due date as well as the ratings after the corrective action(s). An example of a FMEA worksheet is in Appendix F.

Below is a sample application of the DFMEA Worksheet for evaluating the design of a toaster, focusing on the electric cable and its possible failures. The electric cable is only one of many potential issues of the toaster; therefore (in this case) 10 additional worksheets (not shown) are done for the other potential issues. Usually there are many potential issued, all addressed similar to the example shown. Since it is not economical to fix all potential issues, the RPN is calculated to provide a ranking of the issues to provide a guideline where to focus first and which potential issues are not worth spending the effort on. Once the problem is addressed, the RPN is recalculated. If the new RPN fits the requirement of an acceptable managed issue, the team will move on to the next issue or end the FMEA process.

FMEA Worksheet

Page 1 of 11

System, Product, Process: Toaster Design				Owner:		John Doe			Date:		April 1st			
Background				Rating				Countermeasure			Results			
Description	Failure	Effect of Failure	Root Cause	S	O	D	RPN	Action	Who	Due Date	S	O	D	RPN
Electric cable	Insulation fraying	Unreliable Operation or	Incorrect matrial	8	6	2	96	Specify the use of fray resistant	John Doe	1-May				
	Electricity leak		Fatigue	8	4	5	160							
			Wear											
			Corrosion											
			Humidity when coating					Specify tolerable teperature & humidity levels	Jane Doe	3-May				

5. DOE (DESIGN OF EXPERIMENTS)

Design of Experiments is a systematic statistical approach for data collection where variation is present, whether under the full control of the experimenter or not. It consists of planning, conducting, analyzing and interpreting controlled tests to evaluate the factors that control the value of a parameter or group of parameters to be evaluated in the test.

Design of Experiments (DOE) techniques enables designers to determine simultaneously the individual and interactive effects of many factors and their settings (levels) that could affect the output results in any design. It also provides a full insight of interaction between design elements; therefore, helping to turn any standard design into a robust one. Simply put, DOE helps to pin point the sensitive parts and sensitive areas in a design that can cause problems. Designers then are able to fix the problems and produce a robust and higher quality of design prior going into production.

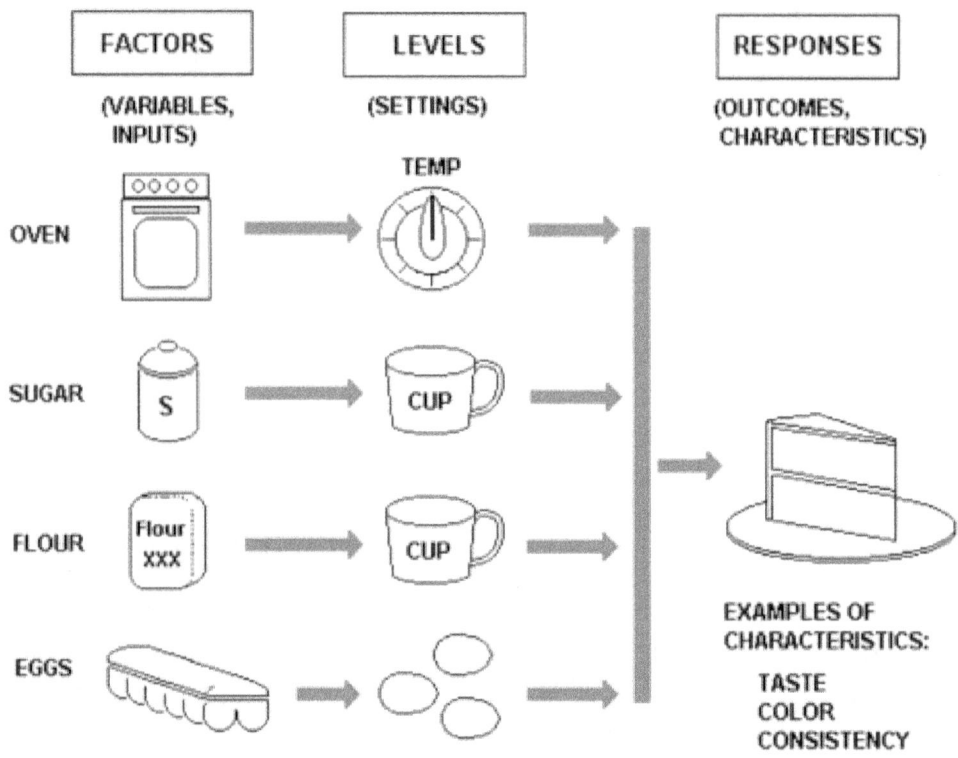

Many conventional experiments involve holding all but one factor constant and altering the levels of another variable. This One–Factor–at–a–Time (or OFAT) approach is inefficient (but easy to understand and explainable by simple statistical models) when compared to changing factor levels simultaneously, which demands more sophisticated statistical models and expertise. Also, OFAT gives no opportunity to evaluate the interdependence of factors, thus delivers not the complete picture.

Most of the current (statistical) DOE are based on taking the time to seriously consider the design and execution of an experiment before executing it. This helps to avoid frequently encountered problems in analysis. Key concepts in creating a designed experiment are:

- **Blocking:** When randomizing a factor is impossible or too costly, blocking lets the designer restrict randomization by carrying out all of the trials with one setting of the one factor and then all the trials with the other settings.
- **Randomization:** Refers to the order in which the trials of an experiment are performed. A randomized sequence helps eliminate effects of unknown or uncontrolled variables (noise).

- **Replication:** Repetition of a complete experimental treatment, including the setup for verification reasons, if feasible.

A well planned and performed experiment should provide answers to questions such as:
- What are the key factors in a process and what is their weight?
- At what settings of the factors would the process deliver acceptable performance?
- What are the key, main and interaction effects in the process?
- What settings would bring less variation in the output?

A good design of experiment typically involves these steps:
1. A screening design which narrows the field of variables under assessment.
2. A "full factorial" design which studies the response of every combination of factors and factor levels, and an attempt to zone in on a region of values where the process is close to optimization.
3. A response surface design to model the response.

The DOE tool provides structured a series of steps to follow when designing experiments and analyzing results. DOE is about understanding and controlling the variation of key process inputs in order to obtain improved results on project outputs. DFSS uses this method for validating and discovering relationships between inputs and outputs.

Elaborating on the example above here is a simple case study of a DOE for the writing of the instructions for a new cake mix. There are four factors to consider and optimize:
1. Ingredients - oil/butter/margarine: 1 (-) versus 2.5 (+) cups
2. Ingredients - eggs: 2 (-) versus 3 (+) eggs
3. Over temperature: 325°F (-) versus 375 °F (+)
4. Cooking time: 25 (-) versus 40 (+) minutes

The high (+) and low (-) points are set using general knowledge of baking; a range that worked with past products well.

In case of using the OFAT approach of try and error and testing every setting between the high and low points, there would be hundreds, maybe even thousands of combination possibilities. Trying this would take a lot of time and money and will most likely lead to only an expectable and not the optimal result.

Using DOE, the experiment will only need 16 batches. It is recommended to always do between 6 to 10 of each setting to account for individual variation.

16 is the number of possible combinations of high and low points for all factors, setting all on high points, all on low points, and all possible combination of it. This is relatively easy to manage. Each resulting cake will be ranked on a scale of 1 to 10 by previously set standards and the results will be entered into a statistics program (Taguchi or Plackett-Burman format). The result of the analysis will pinpoint the interaction of the factors and their optimal settings. The result, the optimal factor setting, may be something like 1.5 cups of oil or butter, 3 eggs, baked at 340 °F for 35 minutes.

A very affordable statistical program for applications like this is IQ Marcos, an add-on tool to Excel.

6. RDO (ROBUST DESIGN OPTIMIZATION)

RDO (Robust Design Optimization) integrates DOE (Design of Experiments) early into the development process, which allows the team to discover optimal solutions to design needs focusing on the minimization of variation. The goal is to reduce variability and increase quality. The solutions developed by RDO will be strong and adaptable enough to accommodate changing focuses and enduring problems.

Variability and tolerance have to be considered for design process of technical systems to assure the highly required quality and reliability. This is the key competency of RDO. There is a class of variability and uncertainty, which is caused by environment influences (temperature, humidity, day light etc.), which can be controlled, but there is also load variation (force, moment), human error etc. which is not controllable. There is also uncertainly, like the weather, which can influence some processes like outdoor plant growth, water supply to a power plant, e.g.. Theses uncontrollable, unpredictable variations cause the uncertainty of satisfaction of the required product functionalities. The design goal is to assure that the specified product functionalities are accomplished even in spite of unavoidable variability and uncertainty in the production process as well as in the use of the product.

There are different statistics based optimization goals:

- Reliability based optimization: Objective is the failure probability of a calculated distribution of the product properties being minimized.

- Variance based optimization: Objective is the variance of a probability distribution being minimized.
- Mean based optimization: Objective is the mean value of a probability distribution being minimized or maximized.

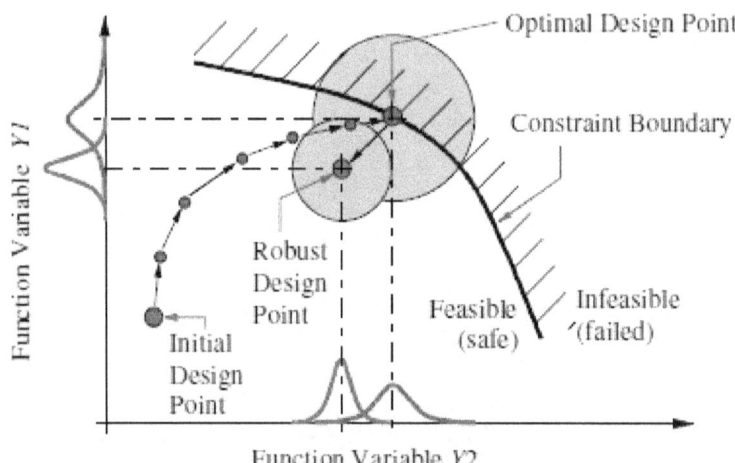

The greatest challenge of the Robust Design Optimization is the long computing time of large product models. Therefore the software and methods should offer a simple integration of key tools (CAD, FEA, cost, post processing, meshing, statistics, etc).

The Robust Design Optimization (RDO) methodology has the following four steps:

- **D**efine:
  - o Understand what is important to the customer and translate into engineering language.
  - o Choose design concepts with variation in mind.
- **C**haracterize:
  - o Generate measurable Critical-to-Quality (CTQ's) criteria for each level.
  - o For each CTQ:
    - Understand the sources of variation.
    - Measure the effects of variation.
- **O**ptimize: For each CTQ choose and implement a strategy to reduce the effects of variation.
- **V**erify: Use knowledge of variation and its effects in construction and design verification plan.

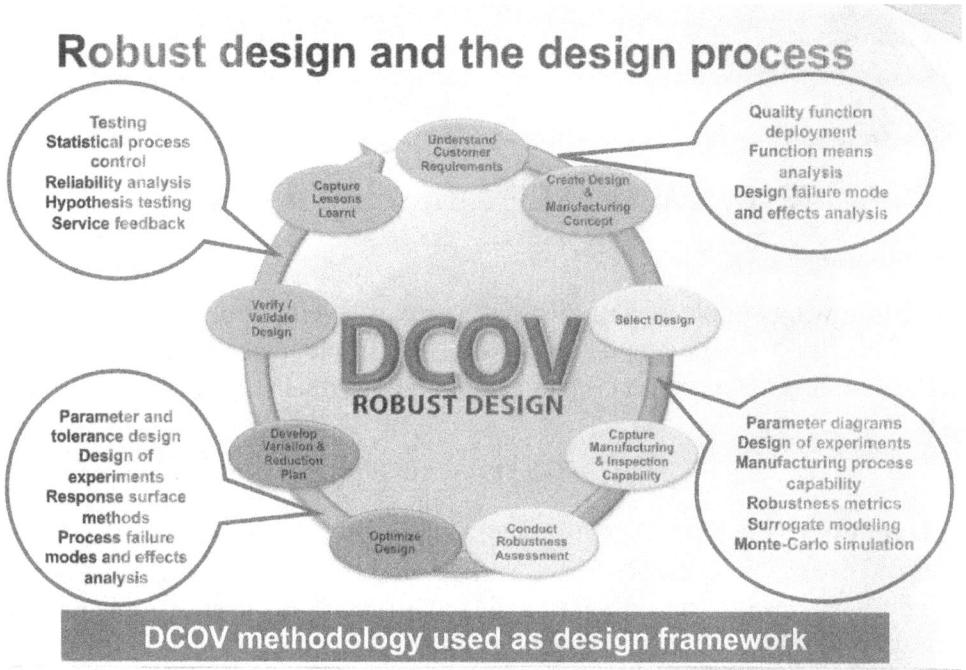

As a result of the implementation of RDO the optimization and automated design helps to achieve better designs by focusing the design work on the key levers. Simulation driven design allows a thorough design assessment early in a development program and it also allows the consideration of variation in the design process and the appropriate selection of the nominal design point; a design that minimizes variation.

APPENDIX A: SIMPLE LEAN DESIGN TOOLS

The following simple Lean tools may help to develop a better process. They should be "no-brainers" and are easy to implement. But sadly, their implementation is not as common as it should be. For this reason I am mentioning them here.

Lean Design

Under lean design I understand the design of a process using simple methods to help speed up the process or prevent errors.

Color coding

One element of lean design is color coding. It's already all over the place. Why not using it too?
Electrical cables are color coded for a reason: blue is "hot", black is the return, and yellow/green is "ground". Red, yellow and green lights have a universal meaning. On your faucet there is a blue symbol for cold and a red symbol for hot water. Everybody knows this, is there is no surprise. The same can be implemented in every other area. My pharmacy puts color coded tags on the medicine bottles, a different one for every household member so they are easily identified. Laboratories still use color codes to identify the process each sample has to go thru (even they now rely more on barcodes). Grease nipples can be color coded to show to which maintenance interval they belong to. Handbooks are color coded to identify different content.

Checklists

Checklists can be a great improvement tool too. Sometimes checklists are very complex, which may have its legitimacy, but in this context I mean short checklists with up to 5 items accompanying the product. Their propose is just to show the advancement in the process so that one can see within a second that's the status is, eliminating the need to read thru some tags or lists. Color coded tags will do a good job too, like triage tags do.

Ergonomics

Ergonomics seems to be the step child of process improvement. It is hardly mentioned, but there is great potential. There are two major areas of ergonomics:

- Workplace ergonomics
- Process flow ergonomics

Workplace ergonomics

The classical workplace ergonomics deals with the ideal height of the chair in relation to the desk or workstation, the ideal placement of the computer monitor and keyboard e.g.. There is plenty of books and article written about this. OSHA in the US and the major health insurance companies in Europe usually provide free information, sometimes even on the job assessments and evaluations or online support.

But there is more to it: 'Soft factors" such as the right lighting and something I call workplace organization. It's part of the 5S element "sort". It is the organization of the workplace in a way that unnecessary tasks or sub-tasks are eliminated. Everything needed should be placed within easy reach. This avoids the unnecessary looking, searching, sorting what can add up in time wasted as well as the unnecessary physical movement associated with it.

Process flow ergonomics

Almost every process has preceding and downstream processes. They all should be in harmony, going hand-in-hand, without delays and long ways to move from one step to the next. They should – but do they?

I remember from my hospital visits when I was young that the ER and Radiology were on the opposite site of the building, preferably even on separate floors. Perfect, since all accident victims in the ER will have to get an X-Ray. That was in the old days. Luckily they now improved even further than expected and have the Radiology on a little card going from patient to patient. A quantum leap improvement; now the equipment comes to the patient instead of the patient being hauled across the building, waiting for hours in another line again.

The floor plan of the modern American house has a process flow / lean perspective too. The garage is next to the kitchen, ideal to move groceries from the car to the refrigerator in an efficient way. Ironically, most families use their garage for everything else but to park the car. But the thought is what counts.

Similar to the workplace ergonomics, process flow ergonomics avoids extra efforts by designing the processes and the lineup of their stations accordingly.

Can we learn from the above examples and redesign our processes for a better process flow?

APPENDIX B: INNOVATION

Against the common notion: **Innovation is a skill, not a talent**. There are a lot of innovative people out there for which it comes natural. But everybody can become an innovator. It is not God given or in somebody's DNA. **It can be learned; it can be trained**.

Here is a short introduction into the matter.

Formal Innovation process

There is a formal process of how innovation is initiated and managed; I call it the corporate way. It consists of brainstorming and evaluation sessions in a formal business meeting setting and is usually run by the Marketing Department. Philip Kottler's "Marketing Management" should be a good source of information regarding this structured process.

A more informal and often in small firms used method is to get some key players together over a case of beer, have a designated "note taker", and start the brainstorming session form there. When I was in the Electronics Industry, this was our way for innovation and we usually came up with pretty good ideas and solution to problems. And it was fun!

Training to be innovative

In order to be an innovative person, one only has to train himself to be innovative. Just going thru the world with open eyes, noticing when something is not optimal and then developing new ways of doing things or better product features are the key. The inspiration can come from something observed as not optimal or just from magazine articles, taking ideas to the next level.

The reason that some people come up with plenty of innovative ideas and others don't is that when somebody it trained on this process, the ideas keep flowing. But if one does not stay within this innovative thought process, the innovative power will decrease. It's like a muscle that needs to get and kept trained.

Innovation and DFSS

DFSS is similar to the formal innovation process mentioned above. It is a standardized process to come up with new (more successful) products, processes, and services; thus innovation. Different as one expected, but by definition, DFSS is Innovation.

APPENDIX C: SEVERITY RATING SCALE

(Should be tailored to meet the needs of the company.)

Rating	Description	Definition (Severity of Effect)
10	Dangerously high	Failure could injure the customer or an employee.
9	Extremely high	Failure would create noncompliance with federal regulations.
8	Very high	Failure renders the unit inoperable or unfit for use.
7	High	Failure causes a high degree of customer dissatisfaction.
6	Moderate	Failure results in a subsystem or partial malfunction of the product.
5	Low	Failure creates enough of a performance loss to cause the customer to complain.
4	Very Low	Failure can be overcome with modifications to the customer's process or product, but there is minor performance loss.
3	Minor	Failure would create a minor nuisance to the customer, but the customer can overcome it without performance loss.
2	Very Minor	Failure may not be readily apparent to the customer, but would have minor effects on the customer's process or product.
1	None	Failure would not be noticeable to the customer and would not affect the customer's process or product.

APPENDIX D: OCCURRENCE RATING SCALE

(Should be tailored to meet the needs of the company.)

Rating	Description	Potential Failure Rate
10	**Very High:** Failure is almost inevitable.	More than one occurrence per day or a probability of more than three occurrences in 10 events ($C_{pk} < 0.33$).
9	**High:** Failures occur almost as often as not.	One occurrence every three to four days or a probability of three occurrences in 10 events ($C_{pk} \approx 0.33$).
8	**High:** Repeated failures.	One occurrence per week or a probability of 5 occurrences in 100 events ($C_{pk} \approx 0.67$).
7	**High:** Failures occur often.	One occurrence every month or one occurrence in 100 events ($C_{pk} \approx 0.83$).
6	**Moderately High:** Frequent failures.	One occurrence every three months or three occurrences in 1,000 events ($C_{pk} \approx 1.00$).
5	**Moderate:** Occasional failures.	One occurrence every six months to one year or five occurrences in 10,000 events ($C_{pk} \approx 1.17$).
4	**Moderately Low:** Infrequent failures.	One occurrence per year or six occurrences in 100,000 events ($C_{pk} \approx 1.33$).

3	**Low:** Relatively few failures.	One occurrence every one to three years or six occurrences in ten million events ($C_{pk} \approx$ 1.67).
2	**Low:** Failures are few and far between.	One occurrence every three to five years or 2 occurrences in one billion events ($C_{pk} \approx$ 2.00).
1	**Remote:** Failure is unlikely.	One occurrence in greater than five years or less than two occurrences in one billion events ($C_{pk} > 2.00$).

APPENDIX E: DETENTION RATING SCALE

(Should be tailored to meet the needs of the company.)

Rating	Description	Definition
10	Absolute Uncertainty	The product is not inspected or the defect caused by failure is not detectable.
9	Very Remote	Product is sampled, inspected, and released based on Acceptable Quality Level (AQL) sampling plans.
8	Remote	Product is accepted based on no defectives in a sample.
7	Very Low	Product is 100% manually inspected in the process.
6	Low	Product is 100% manually inspected using go/no-go or other mistake-proofing gauges.
5	Moderate	Some Statistical Process Control (SPC) is used in process and product is final inspected off-line.
4	Moderately High	SPC is used and there is immediate reaction to out-of-control conditions.
3	High	An effective SPC program is in place with process capabilities (C_{pk}) greater than 1.33.
2	Very High	All product is 100% automatically inspected.
1	Almost Certain	The defect is obvious or there is 100% automatic inspection with regular calibration and preventive maintenance of the inspection equipment.

APPENDIX F: SAMPLE FMEA WORKSHEET

FMEA Worksheet

Description		Countermeasure					Page of

System, Product, Process:

Owner

Date:

Description	Background			Rating				Countermeasure				Results		
	Failure	Effect of Failure	Root Cause	S	O	D	RPN	Action	Who	Due Date	S	O	D	RPN

I

ABOUT THE AUTHOR

Klaus Hogreve was raised and educated in Germany and is currently living in Southern California. He graduated from the University of Lüneburg, Germany, and holds the CMA, CFM, as well as the Six Sigma Black Belt certification.

www.ingramcontent.com/pod-product-compliance
Lightning Source LLC
Chambersburg PA
CBHW080607180526
45168CB00007B/2819